Alice's Adventures in Wonderland

Lewis **Carroll**

Illustrated by **Alida Massari**

Adapted by **Gina D. B. Clemen**

Activities by **Mary Johnson**

Editor: Joanna Burgess
Design and art direction: Nadia Maestri
Computer graphics: Simona Corniola
Picture research: Laura Lagomarsino

First edition : January 2008

Picture credits:
© Cideb Archive; © Bettmann / CORBIS: 4; © Tim Graham / CORBIS: 53.

We would be happy to receive your comments and suggestions, and give you any other information concerning our material.
http://publish.commercialpress.com.hk/blackcat/

ISBN 978 962 07 0448 2 Book + Special CD-ROM

The CD contains an audio section (the recording of the text) and a CD-ROM section (additional games and activities practicing the four skills).

- To listen to the recording, insert the CD into your CD player and it will play as normal. You can also listen to the recording on your computer, by opening your usual CD player program.
- If you put the CD directly into the CD-ROM drive, the software will open automatically.

SYSTEM REQUIREMENTS for CD-ROM	
PC: • Pentium III processor • Windows 98, 2000 or XP • 64 Mb RAM (128Mb RAM recommended) • 800x600 screen resolution 16 bit • 12X CD-ROM drive • Audio card with speakers or headphones	**Macintosh:** • Power PC G3 or above (G4 recommended) • Mac OS 10.1.5 • 128 Mb RAM free for the application
All the trademarks above are copyright.	

Contents

The text is recorded in full.

These symbols indicate the beginning and end of the passages linked to the listening activities.

Lewis Carroll (about 1863)

About the Author

Pen Name: Lewis Carroll

Real Name: Charles Lutwidge Dodgson

Born: 27 January 1832

First book: *Alice's Adventures in Wonderland* (1865)

Other Book: *Through the Looking-Glass* (1871)

Hobby: Photography

Dies: 14 January 1898

The Characters

From left to right, back row: **the Queen, the King.**
From left to right, middle row: **the March Hare, the Hatter, Alice, the Caterpillar.**
From left to right, front row: **the Cheshire Cat, the White Rabbit.**

BEFORE YOU READ

1 VOCABULARY

Here are some words from the beginning of the story. Use them to complete the sentences under the pictures. There is an example at the beginning (0).

pictures doors
key bottle ~~watch~~
cake mouse

0 A rabbit with a ..watch............... .

1 Alice with a
....

2 A box with a
....

3 A cat with a
....

4 A book with
....

5 A hall with
....

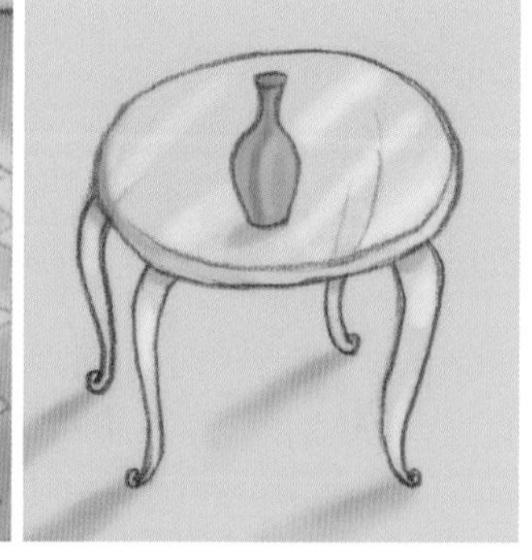

6 A table with a
....

CHAPTER **ONE**

The White Rabbit

It is a sunny afternoon.
Alice and her sister
are sitting by the river.
Alice's sister is reading a book with no pictures.

'I don't like books without any pictures,' Alice thinks.

She is sleepy. Just then she sees a white rabbit.

He looks at his watch and says, 'I'm late!'

'How strange!' Alice thinks. 'A rabbit with a watch!' She follows the rabbit across the grass and into a big hole. She falls slowly down the hole. Then she stops falling and stands up. She looks around.

Alice sees the rabbit again and follows him. He is running around a hall with a lot of doors. Then she sees a glass table with a small key on it. She takes the key and tries to open the doors,

but she can't. Then she sees a very small door and opens it. There is a beautiful garden.

'I want to go into that garden,' thinks Alice, 'but I'm too big.'

She puts the key on the glass table and sees a bottle on it. It says 'DRINK ME' on the bottle. She takes the bottle and drinks it all.

'Oh, how strange!' she says. 'Now I'm very small and I can go into the garden.' But the door is closed and the key is on the table. She is too small now and she can't get the key! She is very sad and starts to cry.

Alice sees a small glass box under the table. Inside the box there is a cake with the words 'EAT ME' on it.

'I'm going to eat it,' says Alice. 'Perhaps I can grow and take the key from the table.'

She eats the cake but nothing happens. Then suddenly she becomes big.

'Now I can get the key,' she thinks.

She takes the key and goes to the door to the garden. But she is too big and can't go in! She sits down and starts to cry again. Her tears [1] make a big pool. [2]

Suddenly she sees the White Rabbit again. He is wearing a lovely jacket and he has got white gloves in one hand and flowers in the other.

'Oh, the Duchess is going to be angry because I'm late,' says the White Rabbit.

'Excuse me, sir...,' says Alice.

The White Rabbit is afraid and runs away. His white gloves fall to the floor. Alice looks at her hand and she is suddenly wearing one of the White Rabbit's gloves.

1. **tears** : these fall from your eyes when you cry.
2. **pool** : an area of water.

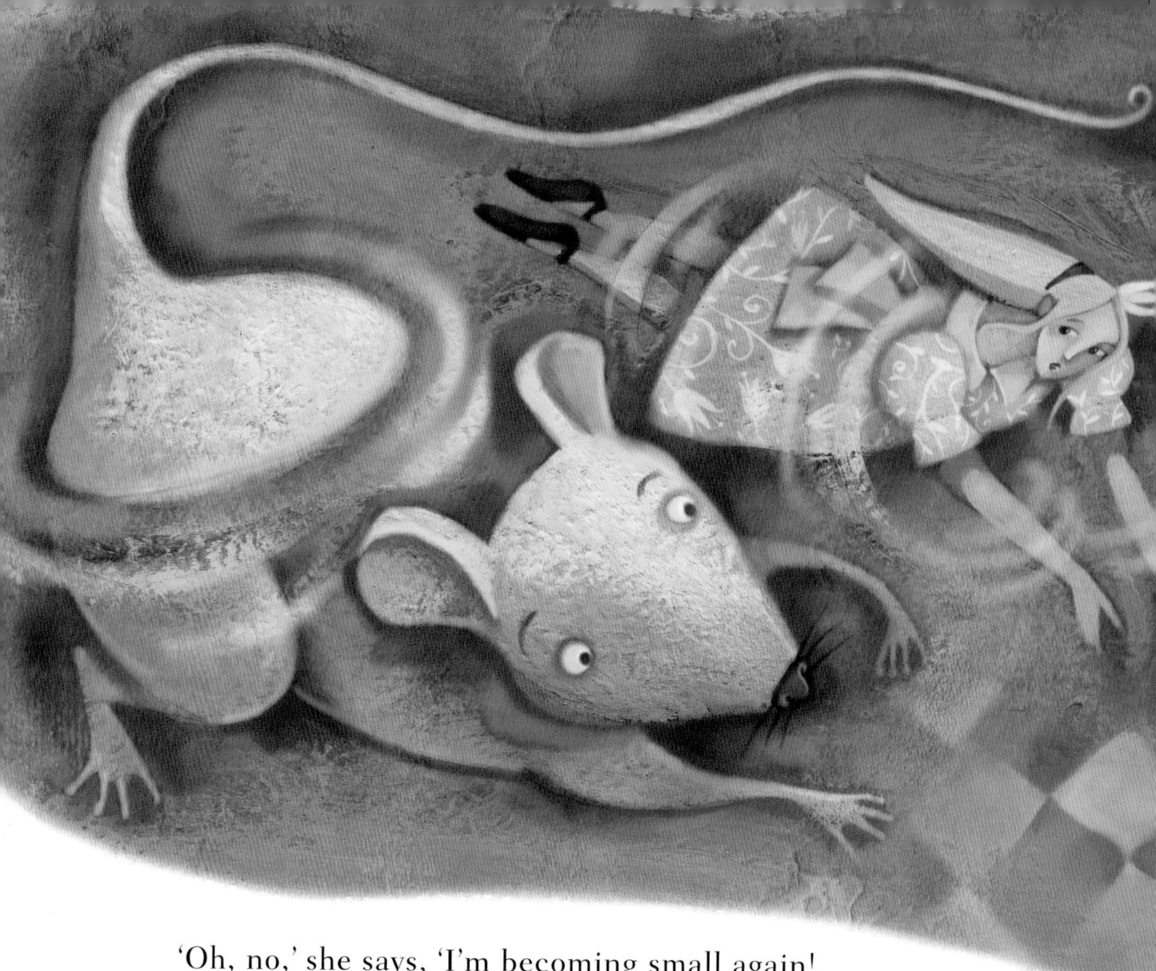

'Oh, no,' she says, 'I'm becoming small again! What's happening to me?'

Alice is small again and she suddenly falls into a pool of water.

'I'm in the sea,' she thinks. But it is not the sea, it is the pool of Alice's tears. She sees a mouse swimming near her.

'Hello, Mouse!' says Alice. 'I'm tired. I want to get out of this pool.'

The Mouse does not answer. 'Perhaps it doesn't understand English,' Alice thinks. 'Maybe it understands French.'

'Où est ma chatte?' [1] asks Alice. This is the first sentence in her French book.

The Mouse is angry and says in English, 'I don't like cats! My family doesn't like cats! I'm a mouse!' He swims away and Alice follows him.

1. **'Où est ma chatte?'** : (French) Where is my cat?

UNDERSTANDING THE TEXT

1 COMPREHENSION CHECK

Choose the best ending to complete these sentences about Chapter One. Tick (✓) A or B. There is an example at the beginning (0).

0 When Alice sees the rabbit
- A ☑ she follows him.
- B ☐ she looks at her watch.

1 When she takes the key
- A ☐ she opens a lot of doors.
- B ☐ she opens a very small door.

2 When she drinks from the bottle
- A ☐ she becomes very small.
- B ☐ she becomes very big.

3 When she eats the cake
- A ☐ she can go in the garden.
- B ☐ she can get the key.

4 When she sees the White Rabbit again
- A ☐ he is wearing white gloves.
- B ☐ he is wearing a jacket.

5 When she meets the Mouse in the pool
- A ☐ she speaks to it in French.
- B ☐ she swims away.

2 'IT SAYS "DRINK ME" ON THE BOTTLE.'

Write the correct phrase under each picture.

Drink me. Eat me. Be quiet. Open me.
Follow me. Read me.

3 ADJECTIVES

A Match the adjectives on the left with their opposites on the right.

1	☐	happy	A	slow
2	☐	big	B	ugly
3	☐	fast	C	sad
4	☐	rich	D	small
5	☐	beautiful	E	new
6	☐	old	F	poor

B Now look at the pictures below and use the words to complete the text. There is an example at the beginning (0).

Tom has got a lot of money. He's very (**0**) ..rich.............. . He lives in a (**1**) .... , (**2**) .... house and he has got a (**3**) .... , (**4**) .... car. But he hasn't got any friends. He's very (**5**) ....

Jerry hasn't got much money. He's very (**6**) .... His car is (**7**) .... and (**8**) .... His house is (**9**) .... and (**10**) .... But he's got a lot of friends. He's very (**11**) ....

4 CONVERSATION

Complete the conversation. What does the White Rabbit say to Alice? Write the correct letter next to the number. There is an example at the beginning (0).

A You must drink from the bottle.

B I haven't got my gloves.

C Yes?

D Then I'm sorry. I can't help you.

E To see the Duchess.

F Because I'm late. I'm very late.

G You mustn't cry.

H No, I can't. I must go.

Alice: Excuse me, sir...

White Rabbit: (**0**) ....

Alice: Can you tell me where I am?

White Rabbit: (**1**) ....

Alice: Where are you going?

White Rabbit: (**2**) ....

Alice: Why are you running?

White Rabbit: (**3**) ....

Alice: Please, wait! How can I get into the garden?

White Rabbit: (**4**) ....

Alice: But the bottle is empty now.

White Rabbit: (**5**) ....

BEFORE YOU READ

1 VOCABULARY

Look at the words below and match them to the pictures.

1	caterpillar	3	duck	5	dodo	7	parrot
2	baby eagle	4	mushroom	6	butterfly	8	pipe

CHAPTER **TWO**

The Blue Caterpillar

Soon there are other strange animals in the pool: a duck, a dodo, a parrot and a baby eagle. Alice gets out of the pool and the animals follow her.

'We're all wet,' says the Dodo. 'Let's run a race and get dry!'

'That's a good idea,' says the Parrot.

Alice and the animals run a race and they get dry.

After the race the Duck asks, 'Who wins this race?'

'Everyone wins and everyone gets a prize,' says the Dodo. 'And Alice is giving the prizes.'

'But I haven't got any prizes,' Alice thinks.

Alice does not know what to do. She puts her hand in her pocket and finds a box of sweets.

'Here are some sweets for everyone,' she says. She gives one sweet to every animal.

'Alice must have a prize too,' says the Mouse.

'Of course,' the Dodo says. 'What have you got for a prize, Alice?'

'I've only got this box,' says Alice and she gives him the empty sweet box.

'Good!' says the Dodo. 'Here's your prize, Alice – a beautiful box.'

'How strange!' she thinks. She looks at the Dodo and says 'Thank you.'

The animals go away and Alice is alone. She hears a noise and sees the White Rabbit.

'The Duchess!' says the White Rabbit. 'She's going to be angry! Where are my gloves?'

Alice looks around. Suddenly everything is different and she is in the countryside.

The White Rabbit sees Alice and says, 'What are you doing here, Mary Ann? Run home and bring me my white gloves!'

'He thinks I'm his servant,' [1] thinks Alice. She runs to the White Rabbit's house and gets his gloves. Then she sees a bottle on a table and thinks, 'Every time I eat or drink something here, interesting things happen.' She drinks it and starts to grow. She is now very big and puts one hand out of the window of the house.

1. **servant** : this person looks after your house.

'Mary Ann! Where are you?' asks the White Rabbit. 'Where are my gloves?'

He tries to open the door of his house. But he can't because Alice's arm is against it.

The White Rabbit calls his gardener. 'Pat! Pat! Where are you?'

'I'm here, sir,' says Pat. There are other animals near the house and they want to help the White Rabbit.

'Come and help me,' says the White Rabbit angrily. 'What's that in the window, Pat?'

'It's a hand,' says Pat.

'A hand!' says the White Rabbit. 'What are you saying? Listen to me. We must burn the house!'

'What!' shouts Alice.

The animals are silent for a moment. Then they throw stones through the windows. The stones become little cakes. Alice eats some of them and becomes small. She is happy and runs out of the house. The animals try to catch her but she runs away into the wood.

Alice sees a big mushroom in the wood. On top of the mushroom there is a sleepy caterpillar. He is smoking a long pipe.

'Who are you?' he asks quietly.

'I... I don't know,' says Alice. 'My size changes all the time. Now I'm very small.'

'I don't understand,' says the Blue Caterpillar.

'I can't explain,' says Alice. 'Let me give you an example; one day you're going to become a butterfly. That's strange, isn't it?'

'No, that's not strange at all,' says the Caterpillar. 'But who are you?'

'I want to know who you are first,' says Alice.

'Why?' asks the Caterpillar.

Alice can't answer the question. She is angry and walks away.

'Come back!' says the Caterpillar. 'I want to tell you something.'

Alice goes back to the Caterpillar and looks at him.

'You must never be angry,' he says.

'Is that all?' asks Alice angrily.

'No,' says the Caterpillar. He smokes his pipe and then gets off the mushroom. 'One side of this mushroom makes you big and the other side makes you small.'

Alice looks at the mushroom and thinks, 'What side of the mushroom have I got to eat?'

She takes a piece from each side and then eats the first piece. Suddenly she grows very small. She eats the other piece and her neck grows long. Then she eats another piece and she becomes the right size. Now she is happy.

UNDERSTANDING THE TEXT

1 COMPREHENSION CHECK

Are these sentences 'Right' (A) or 'Wrong' (B)? If there is not enough information to answer 'Right' (A) or 'Wrong' (B), choose 'Doesn't say' (C). There is an example at the beginning (0).

0 Alice is in the pool with some strange birds.
Ⓐ Right B Wrong C Doesn't say

1 The Duck wins the race.
A Right B Wrong C Doesn't say

2 Alice's prize is different from the other prizes.
A Right B Wrong C Doesn't say

3 The White Rabbit knows Alice's name.
A Right B Wrong C Doesn't say

4 When the animals throw stones Alice is afraid.
A Right B Wrong C Doesn't say

5 Alice thinks it is strange that the caterpillar is smoking a pipe.
A Right B Wrong C Doesn't say

6 After Alice eats two pieces of mushroom she is her normal size again.
A Right B Wrong C Doesn't say

2 THE LIFE CYCLE OF A BUTTERFLY

Look at the pictures. Then put the sentences in the correct order to describe the life cycle of a butterfly. Write the numbers in the boxes.

A ☐ A butterfly comes out.
B ☐ The caterpillar eats and becomes very big.
C ☐ A butterfly lays eggs on a leaf.
D ☐ The caterpillar sleeps and becomes a chrysalis.
E ☐ Baby caterpillars come out of the eggs. They begin to eat.
F ☐ The chrysalis opens.

T: GRADE 2

3 SPEAKING: PETS

There are a lot of animals in Chapter Two. Think of a pet; it can be your pet or a friend's. Tell the class about it. Use these questions to help you.

1 Is this pet yours or a friend's?
2 What animal is he/she?
3 What is his/her name?
4 What colour is he/she?
5 How old is he/she?
6 What does he/she eat?

4 VOCABULARY – FOOD AND DRINK

Reorder the words in the box and use them to label the pictures.

eetwss	ckae	shrumsomo	eat
traew	gubahmrer	gnotuduh	phisc

5 VOCABULARY

Read these definitions of words from Chapter Two. What is the word for each one? The first letter is already there. There is one space for each letter of the word. There is an example at the beginning (0).

0	Alice follows this down the hole.	r a b b i t
1	The Caterpillar is sitting on this.	m _ _ _ _ _ _ _
2	The animals throw these and they become cakes.	s _ _ _ _ _
3	The bottle is on this.	t _ _ _ _
4	The prizes for the animals after the race.	s _ _ _ _ _
5	A colourful bird that can speak.	p _ _ _ _ _
6	The White Rabbit sends Alice to get these.	g _ _ _ _ _
7	The young bird in the pool with Alice.	e _ _ _ _

BEFORE YOU READ

1 LISTENING

Listen to the first part of Chapter Three. For questions 1-5, tick (✓) A, B or C. There is an example at the beginning (0).

0 Alice sees
- A ☑ a nice garden and a small house.
- B ☐ a small garden and a nice house.
- C ☐ a small garden and a small house.

1 Alice is too
- A ☐ tall.
- B ☐ big.
- C ☐ small.

2 The servants have got the faces of
- A ☐ a dog and a fish.
- B ☐ a cat and a fish.
- C ☐ a frog and a fish.

3 The invitation is from
- A ☐ the Queen.
- B ☐ the servant.
- C ☐ the Duchess.

4 The Duchess is sitting
- A ☐ near the cook.
- B ☐ with a cat in her arms.
- C ☐ on a small chair.

5 The cat
- A ☐ doesn't like pepper.
- B ☐ smiles.
- C ☐ is sitting near the Duchess.

CHAPTER **THREE**

The Cheshire Cat

Alice walks in the woods.
She sees a nice garden and a small house.

'I'm too big,' she thinks. 'I can't get into that house. I must eat a piece of mushroom and become small again.' Soon she is nine inches [1] tall.

Suddenly a servant comes out of the wood and goes to the small house. His face is like a fish. Another servant opens the door and his face is like a frog.

The fish-servant has got a big letter in his hand and says, 'For the Duchess. An invitation [2] from the Queen to play croquet.' [3]

The frog-servant says, 'From the Queen! An invitation for the Duchess to play croquet.'

'What strange servants!' says Alice, laughing.

1. **nine inches** : about 23 cm.
2. **invitation** : somebody sends this to ask you to go somewhere or do something.
3. **croquet** : an English game.

Alice goes to the house and says, 'Can I come in?'

'Just open the door and go in,' says the servant.

Then she sees the Duchess. She is sitting on a small chair with a baby in her arms. There is a cook in the kitchen. She is making some soup.

'There's too much pepper in the soup,' Alice thinks and she sneezes. [1]

Then the Duchess sneezes and the baby sneezes.
But the cook and a big cat do not sneeze.
The cat sits near the cook and smiles.

Alice asks the Duchess,
'Why does your cat smile?'

1. **sneezes** :

'Because it's a Cheshire[1] Cat,' says the Duchess. 'All Cheshire Cats smile, don't you know?'

'No, I don't!' says Alice.

'You don't know much,' says the Duchess.

Suddenly the cook starts throwing plates, cups and pots at the Duchess and the baby. There is a terrible noise and Alice is afraid.

'Oh, please be careful!' says Alice. 'The poor baby...!'

'Don't think about the baby,' says the Duchess. 'It's mine!'

She starts singing to the baby, but suddenly she throws it to Alice.

'Here, take the baby!' she says. 'I must go and play croquet with the Queen.' She runs out of the house. The cook throws a plate at her but it does not hit her.

1. **Cheshire** : a part of England.

Alice goes outside with the baby. It makes strange noises. Alice looks at it carefully... it is a baby pig!

'A pig!' says Alice surprised. She puts it down immediately and it runs into the wood.

Just then Alice sees the Cheshire Cat in a tree. It looks at Alice and smiles.

'Hello, Cheshire Cat,' says Alice. 'Where can I go now?'

'Well, where do you want to go?' asks the Cheshire Cat.

'I... I don't know,' says Alice.

'You can go that way on the right and you can visit the Hatter,' says the Cheshire Cat. 'Or you can go that way on the left and you can visit the March Hare. It doesn't matter – they're both mad.'

'Oh dear,' says Alice, 'I don't want to visit mad people.'

'Then you're in the wrong place,' says the Cheshire Cat. 'We're all mad here. You're mad too.'

'How do you know I'm mad?' asks Alice.

'You're here,' says the Cheshire Cat, 'so of course you're mad.'

'Are you going to play croquet with the Queen today?' asks the Cheshire Cat.

'No, I haven't got an invitation,' says Alice.

'I'm going to be there,' says the Cheshire Cat and he suddenly goes away.

Then he comes back and asks, 'Where's the baby?'

'It's not a baby,' says Alice, 'it's a pig!'

'Oh!' he says and goes away again.

Alice goes to the March Hare's house.

'What a big house!' Alice thinks. 'But I'm very small.' She eats another piece of mushroom and she becomes big.

UNDERSTANDING THE TEXT

1 COMPREHENSION CHECK

Are the sentences True (T) or False (F)? Correct the false sentences.

		T	F
1	The Duchess gets an invitation from the Queen.	☐	☐
2	In the house Alice sees the Duchess, a baby, a cook and a cat.	☐	☐
3	The Duchess takes the baby to play croquet with the Queen.	☐	☐
4	Alice doesn't know where to go.	☐	☐
5	The Cheshire Cat doesn't speak to Alice.	☐	☐
6	Alice goes to the Hatter's house.	☐	☐

'SOON SHE IS NINE INCHES TALL.'

In the United Kingdom people still use the traditional 'Imperial measures'.
Here are some Imperial measures and the metric equivalents:

1 inch (in) = 2.5 centimetres (cm) | 1 foot (ft) = 30.5 centimetres
1 yard = 0.9 metres (m) | 1 mile = 1.6 kilometres (km)

2 IMPERIAL MEASURES

Read the sentences and rewrite them using the metric measures in brackets and the word 'about'. You can use a calculator if you want! There is an example at the beginning (0).

0 Big Ben is 320ft tall. (*metres*)
Big Ben is about 96m tall.

1 Edinburgh is 410 miles from London. (*kilometres*)
....

2 The peak of Mount Everest is 29,028ft above sea level. (*metres*)
....

3 Queen Elizabeth is 5ft 4in tall. (*centimetres*)
....

4 My computer screen is 12in wide. (*centimetres*)
....

5 Westminster Abbey is only 200 yards from the Houses of Parliament. (*metres*)
....

3 CRICKET OR CROQUET?

Look at the photos. Cricket and croquet are two traditional British sports. Read the sentences below and decide which sport they are talking about.

A

B

		Cricket A	Croquet B
1	Men, women, old people and young people play this game.	☐	☐
2	The players are usually men.	☐	☐
3	There are four hard balls of different colours.	☐	☐
4	There is one hard red ball.	☐	☐
5	All the players have got a mallet to hit their balls on the grass.	☐	☐
6	One player throws the ball.	☐	☐
7	One player hits the ball with a bat.	☐	☐
8	The players wear special clothes, usually white.	☐	☐
9	The players wear normal clothes.	☐	☐
10	The players hit the balls under six metal hoops.	☐	☐

4 VOCABULARY – COMMON WORD PAIRS

A Use the words in the box to complete these common word pairs.

fork	saucer	chairs	chips	butter	pepper

1 salt and ....
2 table and ....
3 cup and ....
4 knife and ....
5 bread and ....
6 fish and ....

B Match the word pairs to the correct picture.

BEFORE YOU READ

1 VOCABULARY – PLAYING CARDS

Playing cards have got four different symbols on them. These are hearts, spades, diamonds and clubs.
Match the name of the card with the correct picture.

A B C D

1 King of Diamonds
2 Seven of Spades
3 Queen of Hearts
4 Three of Diamonds
5 Jack (Knave) of Spades
6 Ace of Clubs

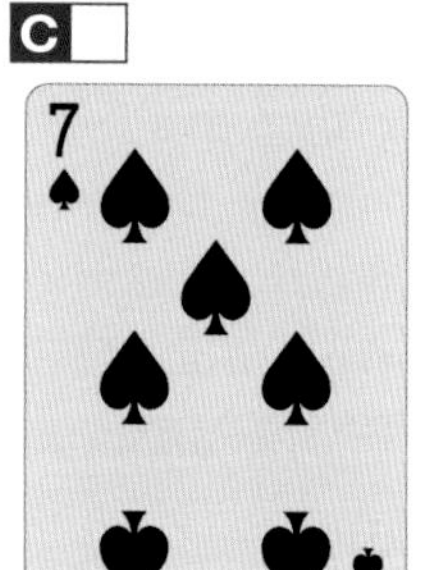

E F

CHAPTER **FOUR**

The Tea Party

There is a big table in front of the house. The March Hare and the Hatter are having tea. A dormouse [1] is on the table between them.

When the March Hare and the Hatter see Alice they say, 'There's no room [2] – no room!'

'But, there's a lot of room,' says Alice and she sits in a big chair.

The Hatter looks at his watch and asks, 'What day is it?'

'I think it's Monday,' says Alice.

'My watch says Wednesday,' says the Hatter.

The March Hare puts his watch in his tea.

Alice is surprised. 'What's he doing?' she thinks.

1. **dormouse** : 

2. **There's no room** : There isn't enough space.

Then the March Hare takes it out and looks at it.

Alice looks at the March Hare's watch and says, 'What a strange watch! It shows the day of the month, but it doesn't show the time.'

'Does your watch show the year?' asks the Hatter.

'Of course not,' says Alice, 'because it's the same year for a long time.'

'Well, it's the same with my watch,' says the Hatter. 'It's always six o'clock here.'

'Have some more tea,' says the March Hare.

'Thank you, but I haven't got any tea,' says Alice. 'How can I have more?

'Oh,' says the Hatter, 'you can always have more than nothing.'

Alice is confused and angry. She gets up from the table and walks away through the trees. The Hatter tries to put the Dormouse into the teapot.

'What a stupid tea party!' she thinks.

Alice is walking in the wood and sees a door in a tree. She opens it and sees the hall with the glass table again. 'This time I want to go into the garden,' she thinks. She takes the key and opens the door. She eats a piece of mushroom and becomes small. Then she walks into the beautiful garden with lovely flowers and fountains. [1]

There is a big rose tree near the door of the garden, and she sees three gardeners. But they are not men – they are playing cards and each of them has got a head, hands and feet. They are painting the white roses red.

1. **fountains** :  they are usually in a park or a square in a town or city.

Alice sees them and says, 'Hello, my name is Alice. Why are you painting the roses red?'

'The Queen hates white roses,' says Five.

'She only likes red roses,' says Seven.

'We must paint them red or she's going to cut off our heads,' says Two.

'Cut off your heads?' says Alice surprised.

Suddenly one of the cards says, 'Look! It's the Queen! The Queen!'

Alice turns around and sees a lot of people. They are all playing cards, each with a head, hands and feet. There are soldiers with clubs, servants with diamonds and spades, children with hearts and Kings and Queens.

UNDERSTANDING THE TEXT

KET

1 COMPREHENSION CHECK

Read these sentences about Chapter Four. Choose the correct answer (A, B or C). There is an example at the beginning (0).

0 The Hatter is sitting
- A ☐ next to the March Hare.
- B ☑ next to the Dormouse.
- C ☐ between the Dormouse and the March Hare.

1 The March Hare puts the watch
- A ☐ in his pocket.
- B ☐ in his tea.
- C ☐ on the table.

2 The Hatter's watch shows the
- A ☐ day.
- B ☐ month.
- C ☐ year.

3 Alice doesn't drink
- A ☐ a lot of tea.
- B ☐ more tea.
- C ☐ any tea.

4 When Alice opens the door in the tree she sees the
- A ☐ garden.
- B ☐ glass table.
- C ☐ mushroom.

5 The gardeners are
- A ☐ men.
- B ☐ servants.
- C ☐ playing cards.

6 They are painting the roses because the Queen likes
- A ☐ red roses.
- B ☐ white roses.
- C ☐ beautiful roses.

2 WORD GAME

Write the days and months in the word game and find the vertical word. The first day of the week is Monday.

1 the third day
2 the eleventh month
3 the first day
4 the seventh day
5 the second day
6 the third month
7 the seventh month
8 the first month
9 the sixth month
10 the sixth day

Vertical word: ....

T: GRADE 2

3 SPEAKING: DAYS AND MONTHS

Ask your partner these questions. Then he/she can ask you!

1 What day is it today?
2 What month is it?
3 When is your birthday?
4 When do you go on holiday?
5 On what days are your English lessons?
6 What is your favourite day of the week?
7 What is your favourite month?
8 On what day do you relax?

'IT SHOWS THE DAY OF THE MONTH BUT IT DOESN'T SHOW THE TIME OF DAY.'

Remember that the verb in the 3rd person singular affirmative takes the '**-s**'.

To make the negative we put **don't** or **doesn't** before the base form of the verb.

*I **don't like** flying.* *He **doesn't live** in the city centre.*

4 PRESENT SIMPLE NEGATIVE

Use the correct affirmative or negative form to make true sentences about the story. There is an example at the beginning (0).

0 Alice ..sits......... (*sit*) in a big chair at the tea party.
1 The March Hare and the Hatter .... (*say*) to Alice: 'There's no room!'
2 The cook and the Cheshire Cat .... (*sneeze*) when the cook puts pepper in the soup.
3 Alice goes away because she .... (*enjoy*) the tea party.
4 The baby .... (*become*) a pig.
5 The Queen .... (*like*) white roses.

5 VOCABULARY

Look at the picture and match the things on the table with the correct words.

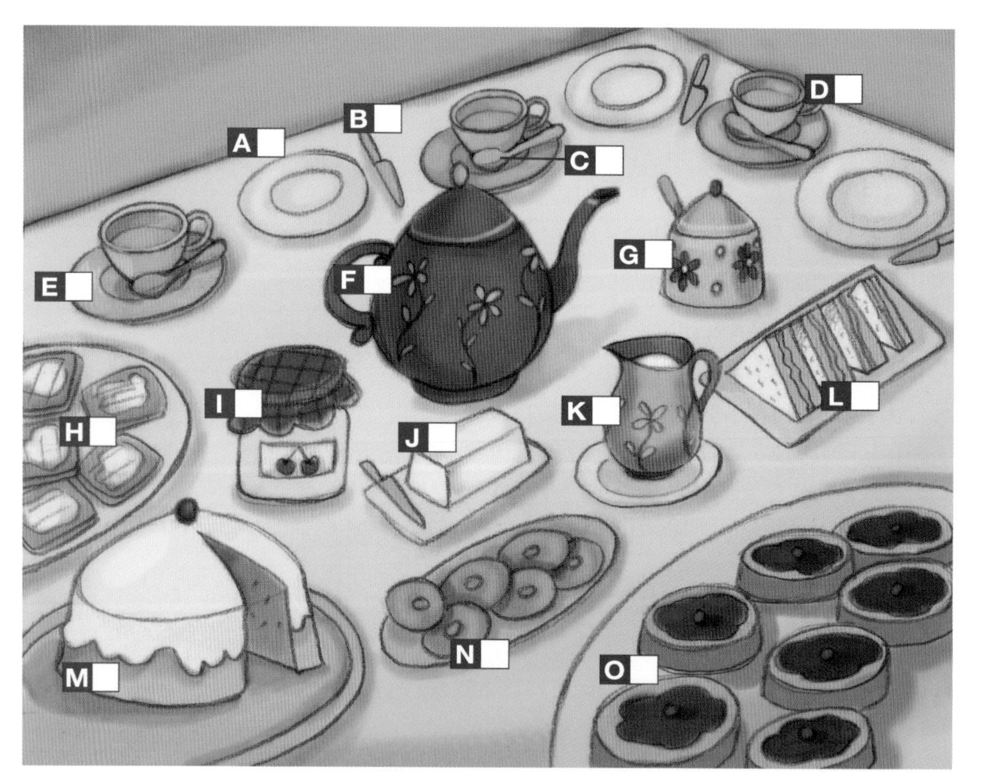

1 plate
2 milk
3 cake
4 sandwiches
5 cup
6 sugar
7 biscuits
8 teaspoon
9 saucer
10 butter
11 tarts
12 teapot
13 jam
14 bread and butter
15 knife

A Drink Called Tea

What is tea?

Tea comes from the dry leaves [1] of the *Camellia sinensis* plant. This plant grows in high places with hot weather and a lot of rain. There are three types of tea: black, green and oolong.

Where does tea come from?

Most tea comes from India, China and Sri Lanka. But some comes from Africa, Malaysia, Indonesia, Japan and South America. Today tea grows in about forty countries. Look at the map and find the countries.

Tea around the world

The British like drinking tea. In the afternoon they sometimes have cakes and sandwiches with it. This is called afternoon tea and it means both the drink and the food, like in the picture on page 38. Some big hotels have got tea rooms where people go to talk and enjoy a cup of tea. You can also get afternoon tea in most cafés.

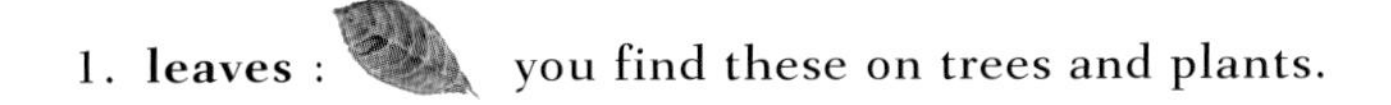
1. **leaves** : you find these on trees and plants.

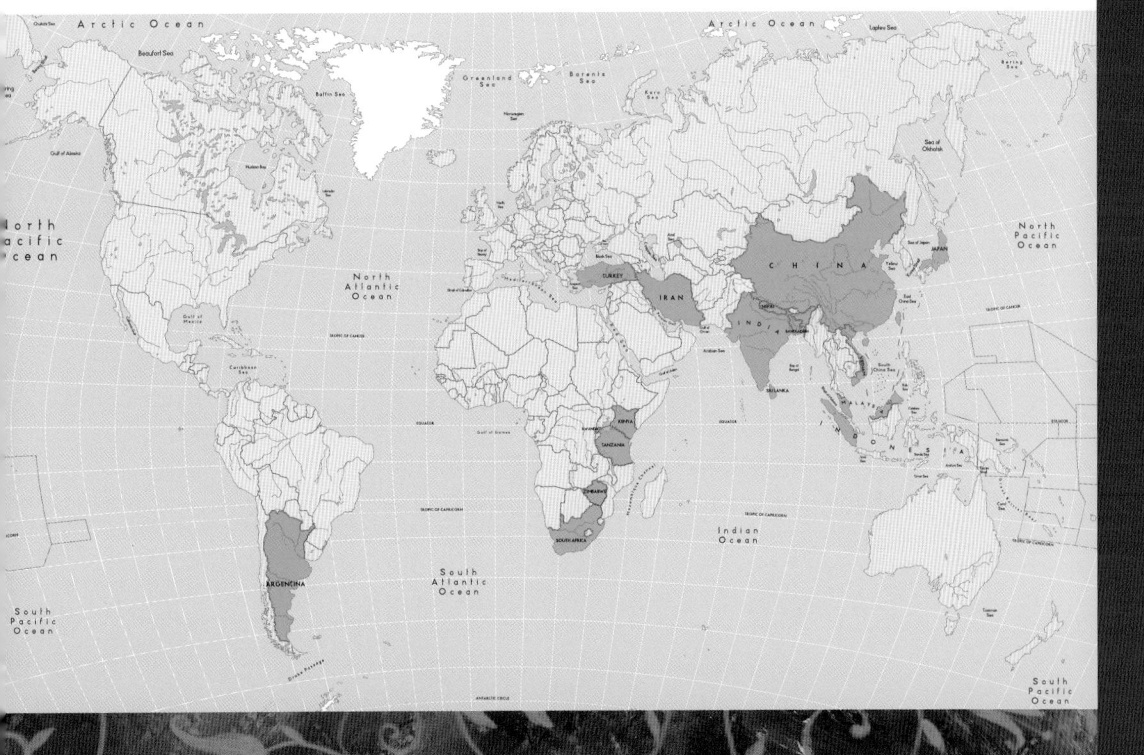

The Japanese have a tea ceremony called Chanoyu. They drink their tea in a special teahouse. Everything is important in this ceremony: the tea, the teapot, the cups and the teaspoons.

People from Turkey drink tea from glasses not cups.

In North Africa and the Middle East people drink strong, hot tea in small glasses with sugar. In some countries of North Africa people drink green tea and mint. [1] It is a good drink in hot weather.

In Turkey people drink tea in special glasses with cubes of sugar. They drink more tea in Turkey than in any other country!

Russian people use a samovar to make tea. A samovar has got a lot of hot water so the Russians can drink tea all day.

You can drink tea hot or cold, with milk or with lemon. A lot of doctors say tea is good for you.

1 COMPREHENSION CHECK

Read the sentences. Match the countries to the correct sentence.

1 Japan
2 North Africa
3 Britain
4 Turkey
5 Russia
6 The Middle East

A ☐ They drink it with cubes of sugar.
B ☐ They drink it with cakes and sandwiches.
C ☐ They drink green tea with mint.
D ☐ They prepare it with a special ceremony.
E ☐ They drink it in small glasses.
F ☐ They drink it in a teahouse.
G ☐ They sometimes drink it in a tea room.
H ☐ They use a samovar to make tea.

1. **mint** : 

CHAPTER **FIVE**

Croquet

The Queen stops and looks at Alice.

'What's your name, child?' asks the Queen.

'My name's Alice,' she says. Then she thinks, 'I mustn't be afraid; they're only cards.'

The Queen looks at the gardeners and asks, 'And who are they?'

'Don't ask me,' says Alice. 'I don't know.'

The Queen is very angry with Alice and says, 'Cut off her head!'

The King looks at the Queen and says, 'But she's only a child, dear.'

The Queen is now angry with the gardeners.

'Cut off their heads!' she says to the soldiers. The gardeners are afraid.

'Alice! Alice!' they shout. 'Please help us!'

'Come here, fast!' says Alice. She puts them in a flower pot [1] and no one can see them.

'Are their heads off?' asks the Queen.

The soldiers can't see the cards.

'Their heads are gone,' say the soldiers.

'That's good,' says the Queen. 'Now let's play croquet! Can you play croquet, dear?'

'Yes!' says Alice.

'Come on, then,' says the Queen. Alice walks away with the Queen, the King and the others.

1. **flower pot** : 

'Go to your places,' shouts the Queen. 'Let's start the game!'

'What a funny game,' Alice thinks. 'The balls are hedgehogs [1] and the mallets are flamingos. [2] It's going to be difficult.'

During the game the Queen is often angry and shouts, 'Cut off his head! Cut off her head!'

'Oh dear,' Alice thinks, 'what's going to happen to my head?'

Suddenly Alice sees the Cheshire Cat. She is happy to see him.

'How are you, dear?' asks the Cheshire Cat.

'I don't like this game,' says Alice. 'No one knows how to play and everyone is angry.'

'Do you like the Queen?' asks the Cheshire Cat.

'No, I don't,' says Alice.

1. **hedgehogs** :  these are small animals.
2. **the mallets are flamingos** : they use these pink birds to play croquet.

The King sees Alice and the Cheshire Cat. 'Who are you talking to?' he asks.

'It's my friend the Cheshire Cat,' says Alice.

'I don't like it,' says the King, 'but it can kiss my hand.'

'No, thank you,' says the Cheshire Cat.

The King is angry and calls the Queen. 'My dear, take this cat away!'

'Of course,' says the Queen, 'cut off its head.' Everyone looks at the Cheshire Cat.

A soldier says, 'I can't cut off its head because it hasn't got a body.'

'It's got a head,' says the King, angrily. 'Cut it off!'

'It's the Duchess's Cheshire Cat,' says Alice. 'Ask her!'

'The Duchess is in prison. [1] Bring her here,' says the Queen to a soldier. Suddenly the Cheshire Cat goes away.

1. **prison** :

UNDERSTANDING THE TEXT

1 COMPREHENSION CHECK

Are these sentences 'Right' (A) or 'Wrong' (B)? If there is not enough information to answer 'Right' (A) or 'Wrong' (B), choose 'Doesn't say' (C). There is an example at the beginning (0).

0 The Queen asks Alice's name.
   (A) Right   B Wrong   C Doesn't say

1 Alice doesn't know the gardeners.
   A Right   B Wrong   C Doesn't say

2 The King wants to cut off Alice's head.
   A Right   B Wrong   C Doesn't say

3 The gardeners are afraid and ask the Queen to help them.
   A Right   B Wrong   C Doesn't say

4 The King likes the gardeners.
   A Right   B Wrong   C Doesn't say

5 The Queen wants to play croquet with Alice.
   A Right   B Wrong   C Doesn't say

6 There aren't real mallets and balls for the game.
   A Right   B Wrong   C Doesn't say

7 Alice likes the Cheshire Cat.
   A Right   B Wrong   C Doesn't say

8 The Duchess takes her cat away.
   A Right   B Wrong   C Doesn't say

2 WRITING

Alice writes to her sister from Wonderland. Complete her letter. Write ONE word for each space. There is an example at the beginning (0).

(0) ...Dear........ Sister,

I (1) .... in a very strange land! There (2) .... a lot of animals and (3) .... can speak! There is (4) .... magic mushroom too. When I eat a piece of the mushroom I become very big or very (5) .... There is a (6) .... who is often angry and she says 'Cut off his (7) ....!' I think (8) .... is a little mad.

I hope I (9) .... come home soon. Say hello to Mother (10) .... me.

Love,
Alice

3 A GAME OF CROQUET

Join all the croquet balls in the correct order to make a sentence. You must go under all the hoops.
When you finish what can you see? A _ _ _ _ _

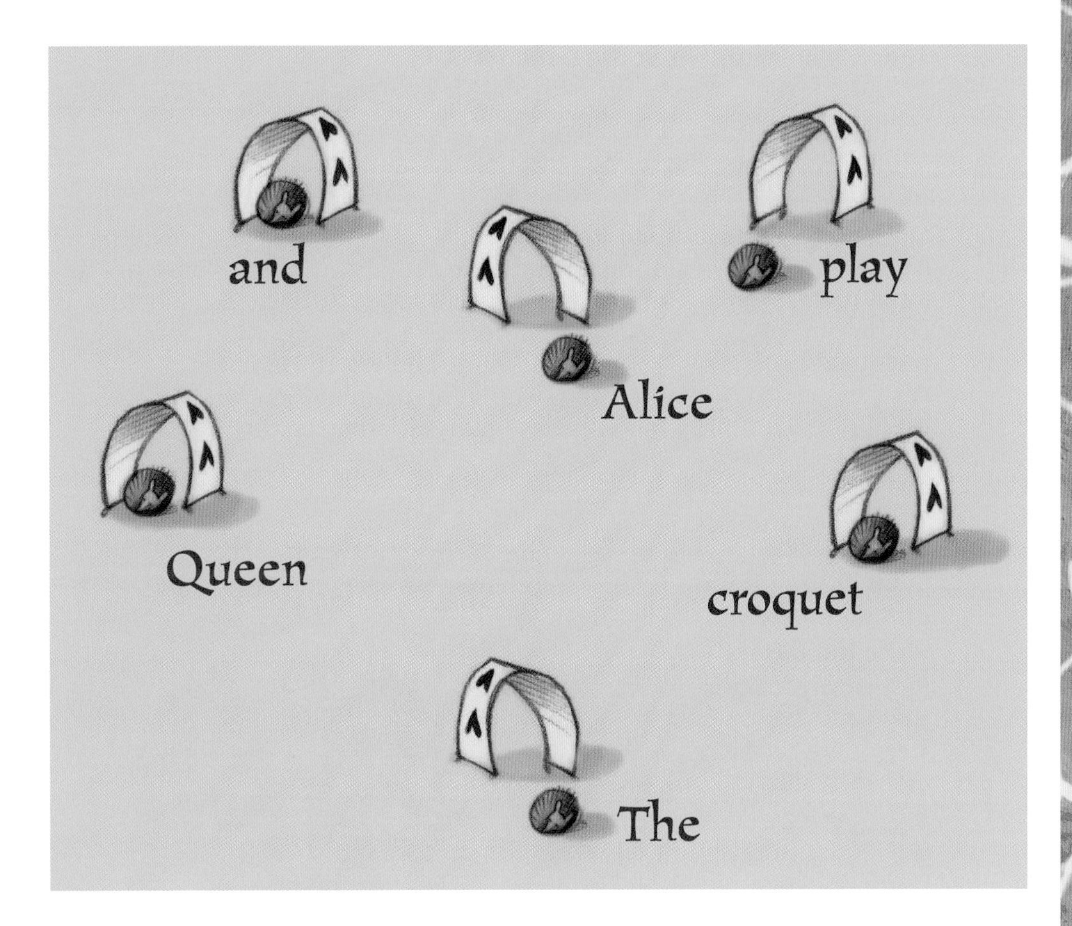

4 SPEAKING – SPORT

Prepare a short talk about a sport. Use these questions to help you. Use a dictionary for new words.

1 What is the name of the sport?
2 Is it an individual sport or are there teams?
3 Where do people do this sport?
4 What equipment do people use?
5 What special clothes do they wear?
6 Can people do this sport in all months of the year?

BEFORE YOU READ

1 LISTENING

A Listen to the first part of Chapter Six. It takes place in a courtroom (see picture below). Fill in the gaps with the words from the box. There is an example at the beginning (0).

near	in	in the middle	in front of	with	on

She is soon (0) ...in..................... the courtroom. The King and the Queen of Hearts are sitting (1) .... their thrones. The Knave of Hearts is standing (2) .... them. There are a lot of birds, animals and cards sitting (3) .... the courtroom. (4) .... there is a table (5) .... a plate of tarts (6) .... it. The White Rabbit is (7) .... the King, with a long piece of paper (8) .... his hand. He starts reading.

B Listen again. Look at the picture of the courtroom. Write the letters in the correct boxes.

1 ☐ the Knave of Hearts
2 ☐ the Queen
3 ☐ the White Rabbit
4 ☐ plate of tarts
5 ☐ birds, animals and cards
6 ☐ the King

CHAPTER **SIX**

The Trial [1]

Suddenly someone shouts, 'The trial's beginning! The trial's beginning!'

'A trial?' asks Alice. 'Whose trial is it?' But no one answers Alice.

She is soon in the courtroom. The King and Queen of Hearts are sitting on their thrones. The Knave of Hearts [2] is standing in front of them. There are a lot of birds, animals and cards sitting in the courtroom. In the middle there is a table with a big plate of tarts on it.

The White Rabbit is near the King with a long piece of paper in his hand. He starts reading:

'The Queen of Hearts, she makes some tarts
All on a summer day.
But the Knave of Hearts,
... He takes the tarts away!'

1. **Trial** : a meeting in a courtroom.
2. **Knave of Hearts** : 

'Cut off his head!' shouts the Queen.

'No, no!' says the White Rabbit. 'We've got to listen to some witnesses [1] first.'

'Very well,' says the King. 'Call the first witness.'

The first witness is the Hatter. He has got a teacup in one hand and bread and butter in the other.

'I'm sorry about this but it's my tea time,' says the Hatter.

'Oh, really?' says the King. 'Take off your hat!'

'It isn't mine,' says the Hatter.

'Whose hat is it, then?' asks the King angrily.

1. **witnesses** : these people give information at a trial.

'I don't know,' says the Hatter. 'I sell hats.'

'Tell me what you know,' says the King.

'Oh, I don't know anything,' says the Hatter.

The Queen looks at the Hatter with her glasses. He is afraid of her and his face is white. He bites a piece of his teacup instead of his bread and butter.

'I'm just a poor man...,' says the Hatter sadly. 'Please let me go and finish my tea.'

'Very well, you can go,' says the King.

'Cut off his head outside,' the Queen says to one of the soldiers. But the Hatter runs away quickly and no one can catch him.

Suddenly Alice feels strange. 'Oh no,' she thinks, 'I'm growing again.'

'Call the next witness,' shouts the King.

The next witness is the Duchess's cook. She comes into the courtroom with a big pepper pot and everyone sneezes.

'Tell me everything you know,' says the King.

'No!' says the cook.

The King is surprised and looks at the White Rabbit.

'Ask the witness questions, your Majesty,' says the White Rabbit quietly.

'Oh, that's right,' says the King. 'What are tarts made of?'

'They're made of pepper,' answers the cook.

'They're made of sugar,' says the sleepy Dormouse.

'What!' says the Queen. 'Send the Dormouse away – cut off its head!'

There is a lot of noise in the courtroom and at last the Dormouse goes away.

'Call the next witness,' says the King.

The White Rabbit looks at his long piece of paper and says, 'Alice!'

Alice is very surprised. She is quite big now. She stands up quickly and some of the birds and animals fall over.

'Oh,' says Alice, 'I'm very sorry!'

Then she goes and stands in front of the King and Queen.

'What do you know about this?' asks the King.

'Nothing,' answers Alice.

'That's very important,' says the King.

'You mean unimportant, your Majesty,' says the White Rabbit.

'Of course,' says the King, 'I mean... unimportant.' Then he writes something in a book.

UNDERSTANDING THE TEXT

1 COMPREHENSION CHECK

Match the questions (1-8) with the correct answer (A-H).

1 ☐ Where is the trial?
2 ☐ Who is the first witness?
3 ☐ Why does everyone sneeze?
4 ☐ What does the dormouse say?
5 ☐ When does the White Rabbit call Alice?
6 ☐ What does Alice know?
7 ☐ How many witnesses are there?
8 ☐ What has the Hatter got in one hand?

A a teacup
B nothing
C three
D in the courtroom
E the Hatter
F that tarts are made of sugar
G after the dormouse goes away
H because the cook has got a big pepper pot

2 MAKE A CUP OF ENGLISH TEA!

Read the instructions and number the pictures in the correct order.

1 Boil some water.
2 Pour some water into the teapot to make it warm. Then pour it out.
3 Put 2 or 3 teabags into the teapot.
4 Pour the boiling water into the teapot.
5 Wait for 5 minutes.
6 Drink the tea with a little milk, and a little sugar if you like it!

A Real Queen

- Her name is Elizabeth Alexandra Mary Windsor.
- Her date of birth is 21 April 1926.
- She becomes Queen on 27 May 1952.
- Her title is Queen Elizabeth II.
- She is Queen of The United Kingdom of Great Britain and Northern Ireland.
- Her husband is Philip, Duke of Edinburgh.
- She has got three sons (Charles, Andrew and Edward) and one daughter (Anne).
- She has got eight grandchildren.
- She loves dogs, especially corgis.
- She loves horses and she loves horse riding.
- When she is in London she lives in Buckingham Palace.
- Her other homes are Windsor Castle, near London and Sandringham House, in the east of England.
- She has got a castle in Scotland, Balmoral Castle. She goes here for her summer holidays.

1 COMPREHENSION CHECK
Now answer the following questions.

1 What is the Queen's family name?
2 What is her husband's title?
3 How many children has she got?
4 Which animals does she love?
5 What is her favourite sport?
6 What's the name of her home in London?
7 What's the name of her holiday home in Scotland?
8 How old is she?

PROJECT ON THE WEB

LET'S FIND OUT MORE ABOUT THE QUEEN'S FAMILY.

Connect to the Internet and go to www.blackcat-cideb.com or www.cideb.it. Insert the title or part of the title of the book into our search engine. Open the page for *Alice's Adventures in Wonderland.* Click on the Internet project link for the Royal Family.

Find more information about the British royal family.

A Write the names in the Queen's family tree.

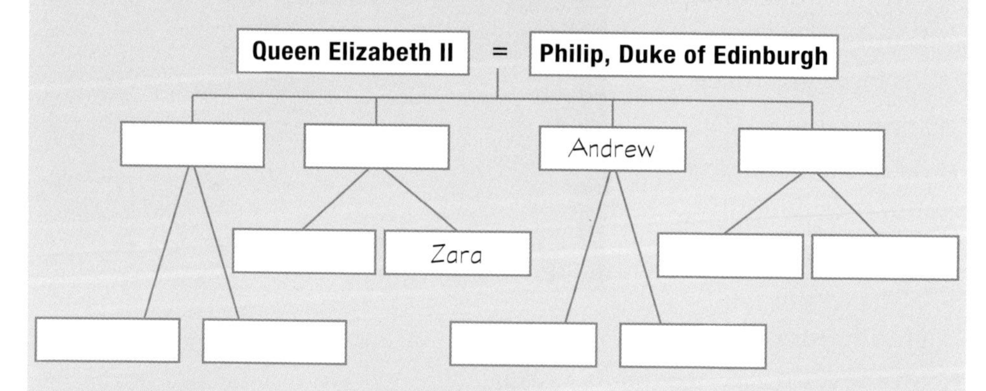

B Find the answers to these questions.

1 How many children has the Princess Royal got?
2 What is the name of Edward's first child?
3 Who is Prince Charles's second son?
4 What is the name of Zara Phillips's brother?
5 Who has got two daughters?
6 Who is Prince Charles's wife now?
7 Who is Eugenie's sister?
8 Who is the Queen's last grandchild?

CHAPTER **SEVEN**

Back Home for Tea

10

'Silence in the courtroom!' shouts the King. 'I have got an important rule[1] for you.'

Everyone is silent and looks at the King. He opens a book and reads, 'Rule Forty-two. All people more than a mile tall must leave the courtroom.'

Everyone looks at Alice.

'I'm not a mile tall,' says Alice.

'Yes, you are!' says the King.

'You're more than two miles tall,' says the Queen.

'I don't want to leave the courtroom,' says Alice. 'It's not a real rule.'

'It's a very old rule,' says the King.

'Then why is it Rule Forty-two and not Rule One?' asks Alice.

The King does not know what to say and closes the book quickly.

The White Rabbit jumps up and says, 'Look, I've got a letter!' He puts his glasses on and looks at it.

1. **rule** : a rule says what you must do.

'Oh! It's not a letter, it's a poem,' he says.

'Read it from the beginning and stop at the end,' says the King.

The White Rabbit reads the poem but no one understands it.

'This poem is nonsense,' [1] says Alice.

'Oh, cut off the Knave's head!' says the Queen.

'What nonsense!' shouts Alice. She is not afraid of anyone now because she is very big.

'Be quiet!' shouts the Queen.

'No!' shouts Alice.

The Queen is very angry and her face is purple.

'Cut off her head!' she shouts.

'I'm not afraid of you,' says Alice. 'You're only cards!'

Suddenly the cards fly up into the air and fall down on her.

1. **nonsense** : silly, stupid.

'Oh, dear!' says Alice. She pushes the cards away from her face.

Alice wakes up. These cards are leaves! Her sister pushes them away from her face.

'Wake up, Alice dear!' says her sister.

'Oh, what a strange dream!' [1] says Alice.

'Please tell me about it,' says her sister.

Alice tells her sister about the dream.

Her sister listens and then laughs.

'Yes,' she says, 'it's a very strange dream. But it's late now and it's time for your tea.'

Alice runs home and thinks about the White Rabbit, the Caterpillar, the Duchess, the Hatter, the Cheshire Cat, the croquet game, the Queen, the King and the cards.

'What a wonderful dream!' she says happily. 'One day I can tell my children about it.'

1. **dream** : what you think when you are asleep.

UNDERSTANDING THE TEXT

1 COMPREHENSION CHECK

Read the summary of Chapter Seven. Choose the best word (A, B or C) for each space to say what happened in this chapter. There is an example at the beginning (0).

The King says that Alice (0) ..must............ leave the courtroom because she is very (1) .... Alice (2) .... want to leave. The White Rabbit begins to (3) .... a poem but no one (4) .... understand it. The Queen wants to cut (5) .... the Knave's head and she is very (6) .... when Alice says 'No!'

Suddenly the cards all fly in the air and fall on Alice's (7) .... At that moment she wakes up and sees that the cards are leaves. She is with her sister again. What a (8) .... dream!

	A	B	C
0	(A) must	B mustn't	C can't
1	A small	B tall	C silent
2	A don't	B doesn't	C isn't
3	A sing	B write	C read
4	A can	B must	C wants
5	A on	B off	C up
6	A quiet	B happy	C angry
7	A face	B hands	C feet
8	A horrible	B strange	C stupid

2 SPEAKING – FREE TIME

Alice has her adventures in Wonderland on a sunny afternoon. She is sitting by the river with her sister. Tell your partner about your favourite activities on a sunny afternoon in the summer.

1. Where do you usually go?
2. Who do you go there with?
3. Do you talk, play, read, swim...? What do you usually do?
4. What do you usually wear?
5. What time do you usually go home?

3 NOTICES

Which notice (A-H) says this (1-5)? There is an example at the beginning (0).

A

Special offer
Cream teas!
£ 3.00 for two.
Sat and Sun only.

B

Jam tarts!
Buy one get one free!
Only 95p for 2.

C

Wonderland Tour!
Tickets at kiosk.
Children £1 Adults £3.
Every weekend.

D

Please
be silent
at all times.

E

No animals
allowed in the
play area.

F

G

THESE TARTS
DO NOT CONTAIN
SUGAR

H

0 ..H... You can go in if you are 1.20m tall.
1 ...... You can buy two for the price of one.
2 ...... Mothers and fathers pay £3.
3 ...... You can eat these if you are on a diet.
4 ...... At the weekend two people pay only £3.
5 ...... You can't take your dog here.

4 PICTURE SUMMARY

Put the pictures in the correct order to retell the story. Then write a sentence to describe each picture.

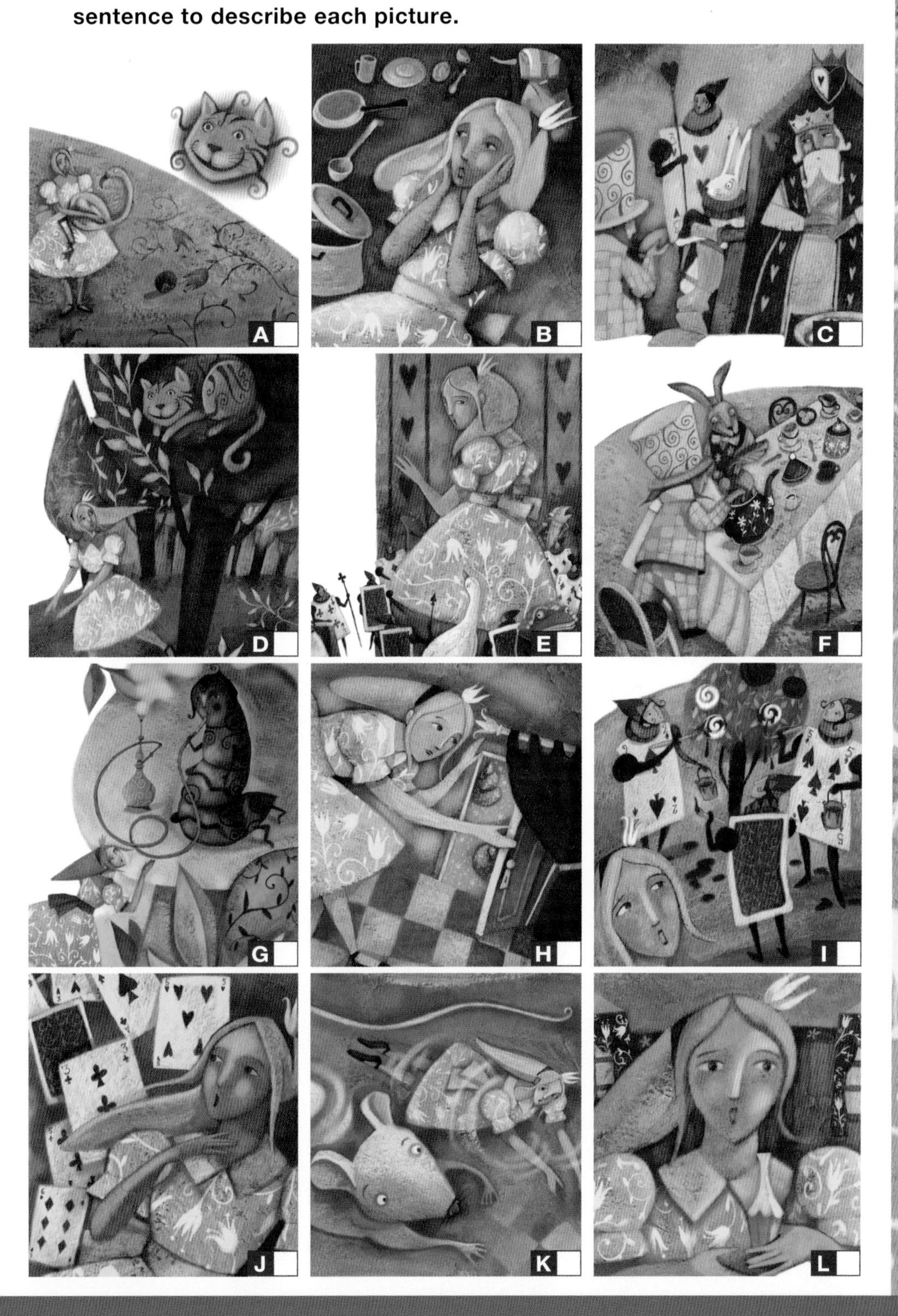

EXIT TEST

1 COMPREHENSION CHECK

Read the sentences about the story. Decide if the sentences are true (T) or false (F).

		T	F
1	Alice falls into a hole because she follows her sister.		
2	She finds a bottle with the notice 'DRINK ME'.		
3	She finds a mushroom with the notice 'EAT ME'.		
4	She becomes big and small when she eats and drinks.		
5	The Caterpillar doesn't speak to Alice.		
6	With the Duchess are a cook, a cat and a baby.		
7	The baby becomes a pig.		
8	At the March Hare's tea party are the Dormouse and the Mad Hatter.		
9	Alice has a very normal game of croquet with the Queen of Hearts.		
10	The Knave is in the courtroom because he takes the Queen's tarts.		
11	Alice leaves the courtroom because she is one mile tall.		
12	Alice wakes up and understands that her adventure is a dream.		

SCORE

2 CROSSWORD

Complete the crossword.

1

2 We wear them on our hands in the winter.

3

4 Alice drinks from it and she becomes very small.

5 It tells us the time.

6 Alice sees a blue one. It is smoking a pipe.

7

SCORE

3 WHO SAYS WHAT?

Match each sentence (1-8) with the character who says it (A-H).

A	Alice	C	the Queen	E	the White Rabbit	G	the Hatter
B	the Caterpillar	D	the gardeners	F	the Cheshire Cat	H	the King

1 ☐ 'Silence in the courtroom!'
2 ☐ 'Cut off her head!'
3 ☐ 'The Queen hates white roses.'
4 ☐ 'It's always six o'clock here.'
5 ☐ 'We're all mad here. You're mad too.'
6 ☐ 'The Duchess is going to be angry because I'm late.'
7 ☐ 'Why does your cat smile?'
8 ☐ 'One side of this mushroom makes you big and the other makes you small.'

SCORE

4 HOW MUCH CAN YOU REMEMBER?

Answer the following questions.

1 What does the Mad Hatter drink?
2 What does Alice find inside a small glass box?
3 What does Alice meet when she swims in the pool of tears?
4 What game do Alice and the Queen play with flamingos and hedgehogs?
5 What does the cook put in the soup?
6 What does the Queen send to the Duchess?
7 What do the King and Queen sit on in the courtroom?
8 What does the Caterpillar smoke?
9 What does Alice follow into the hole at the beginning of the story?
10 What does Alice try to open with a key?

SCORE

KEY TO EXIT TEST

1 **1** F; **2** T; **3** F; **4** T; **5** F; **6** T; **7** T; **8** T; **9** F; **10** T; **11** F; **12** T.

2 **1** fountain; **2** gloves; **3** prison; **4** bottle; **5** watch; **6** caterpillar; **7** pig.

3 **1** H; **2** C; **3** D; **4** G; **5** F; **6** E; **7** A; **8** B.

4 **1** tea; **2** a cake; **3** a mouse (and other strange animals); **4** croquet; **5** pepper; **6** an invitation; **7** thrones; **8** a pipe; **9** the White Rabbit; **10** the door to the garden.

This reader uses the **EXPANSIVE READING** approach, where the text becomes a springboard to improve language skills and to explore historical background, cultural connections and other topics suggested by the text.
The new structures introduced in this step of our **G**REEN **A**PPLE series are listed below. Naturally, structures from lower steps are included too.
The vocabulary used at each step is carefully checked against vocabulary lists used for internationally recognised examinations.

Starter A1

Verb tenses
Present Simple
Present Continuous
Future reference: Present Continuous; *going to*; Present Simple

Verb forms and patterns
Affirmative, negative, interrogative
Short answers
Imperative: 2nd person; *let's*
Infinitives after some very common verbs (e.g. *want*)
Gerunds (verb + *-ing*) after some very common verbs (e.g. *like*, *hate*)

Modal verbs
Can: ability; requests; permission
Would ... like: offers, requests
Shall: suggestions; offers
Must: personal obligation
Have (got) to: external obligation
Need: necessity

Types of clause
Co-ordination: *but*; *and*; *or*; *and then*
Subordination (in the Present Simple or Present Continuous) after verbs such as: *to be sure*; *to know*; *to think*; *to believe*; *to hope*, *to say*; *to tell*
Subordination after: *because*, *when*

Other
Zero, definite and indefinite articles
Possessive *'s* and *s'*
Countable and uncountable nouns
Some, any; *much, many, a lot*; *(a) little, (a) few*; *all, every*; etc.
Order of adjectives

Available:

- **Peter Pan**
  J. M. Barrie
- **Alice's Adventures in Wonderland**
  Lewis Carroll
- **The Wonderful Wizard of Oz**
  L. Frank Baum
- **The Happy Prince and The Selfish Giant**
  Oscar Wilde
- **The Adventures of Tom Sawyer**
  Mark Twain
- **The Secret Garden**
  Frances Hodgson Burnett
- **Sherlock Holmes Investigates**
  Sir Arthur Conan Doyle
- **Around the World in Eighty Days**
  Jules Verne
- **Romeo and Juliet**
  William Shakespeare

Starter A1